People create paper, and
Paper creates culture.
People protect the forest, and
People are protected by the forest.

人は紙をつくり

紙は文化をつくる。

人は森を守り

人は森に守られる。

目次

登場人物（Characters）

ショータ（S）

ワカミ（W）

グランパ（G）

川上御前（K）

Contents

Chapter 1: What is Washi (Japanese Paper)?
Chapter 2: The Washi-making Process (Kozo paper)
Chapter 3: Japanese History and Washi
Chapter 4: More information about Washi

第1章 和紙ってなに？

What is Washi (Japanese Paper)?

❶ S Hi, I'm Shohta, I'm in the 5th grade, I'm very good at baseball but not at my studies…

❷ S Today, my parents gave me a new smartphone as I got high marks in my exams! I will show it to my friends!

❸ S Here is my new smartphone everyone!

❹ S Oh, No! No! Just look, guys!

❺ S I better get a new cover ASAP. I completely broke my old one.

ワカミちゃんのカバー、イケてるじゃん。それどこで買った?

私のおじいちゃんのこと。紙でできてるの、コレ。

もらったの。誕生日にグランパから。

ぐ、ぐらんぱ?

⑥ Ⓦ It's a birthday present from Grandpa. Ⓢ Grandpa?

⑦ Ⓦ This is made of paper.

Ⓢ How nice your cover is, Wakami! Where did you buy it? Amaxon?

え!紙い?じゃあ雨とか、濡らすとダメだろ?

オレ、汗かきで、雨男だし。

⑧ Ⓢ What? Paper? It must be weak in water then? I'm afraid that it's not suitable for me…

軽くて丈夫?スグレモノじゃん。

でもマジか?

ぜ〜んぜん。それに、紙だから軽いし。

⑩ Ⓢ Light and durable? That's nice! But, really?

⑨ Ⓦ Not at all. It is also light.

② Ⓢ 1 year…Hmm…is paper so long-lasting?

① Ⓦ Yeah, I have been using this since last year.

④ Ⓦ It's "Washi" according to Grandpa!

③ Ⓦ This isn't ordinary paper. It's Washi.
③ Ⓢ What?

Ⓢ "Can I wash my cup, my dear?"
⑤ Ⓦ "Not "Wash"! But "WASHI"!"

ワシは
和紙を
よう
知らん。

いい加減に
してよ

Ⓢ "Oh…I don't know how to wash Washi."
⑥ Ⓦ Enough.

ちょっと変な人だけど。

よかったら帰りに
グランパの家へ
一緒に行く？
和紙好きで、他にも
いろいろ集めてるよ。

Ⓦ Do you wanna go to Grandpa's house?
He is a bit of a strange man, collecting
⑦ various Washi items.

Ⓢ OK! So, let's go
now!
Make hey while
the sunshine!
Ⓦ It's "make 'hay'";
and wait a
minute.
⑧

オーライ！
では早速。

「変は急げ」
って言うし。

「善」でしょ。

今すぐ？
ま、まだ
説明が…

ハウ
ドゥ
ユドゥー、
グランパ！

Oh！
ボーイ
フレンド？
留学生か？

どっちも
違うって

Ⓢ How do you do Grandpa?
Ⓖ Oh! Your boyfriend? Or an overseas student?
⑩ Ⓦ Neither.

グランパ、
いるー？

Oh！
ワカミ？

Ⓦ Hi, Grandpa!
⑨ Ⓖ Oh, Wakami?

② Ⓢ Whaaaaat? Ⓦ Grandpa is British.

Ⓢ Oh...He, Hello, Mai neimu iz Shohta...
③ Ⓖ Oh, Shorter?

① Ⓖ Fine, thanks for asking. Nice to meet you.

⑤ Ⓦ He likes the smartphone case you gave to me. So please tell him about Washi.

④ Ⓦ Don't be mean to this dumb bunny. Don't worry. He speaks Japanese well.

⑦ Ⓖ That's right! Washi has been beautiful and durable from the year 'dot'.

⑥ Ⓖ Oh, the one made of Washi! Why did you like it?
Ⓢ It has beautiful colors, and is water-resistant and durable…

⑧ Ⓢ From the year dot? Meiji Era or Edo Period?
Ⓚ No…from more ancient times.

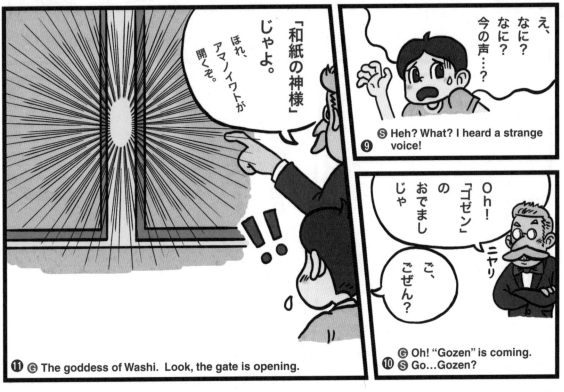

⑨ Ⓢ Heh? What? I heard a strange voice!

⑩ Ⓖ Oh! "Gozen" is coming.
Ⓢ Go…Gozen?

⑪ Ⓖ The goddess of Washi. Look, the gate is opening.

K TA-DA!!!!
① S What? Someone appeared out of nowhere!

S Y..Yes, but are you the goddess of Washi? Your name?
③ G She is more like a fairy rather than a goddess. S "Fairy?"

② K You want to know the world of Washi, right?

G It's the origin of Washi. About 1500 years ago…
⑥ S 1500 years ago!? That's when people made 'tumuli', isn't it? I know history.

S Oh, she has disappeared!
⑤ G Just like the legend! Always exciting!
S What's "the legend"?

K "I'm the one who lives in the upper stream of Okamoto River…"
④ (In Fukui)

その通り！

わ、また
現れた!?

Ⓚ Boink! That's correct!
⑦ Ⓢ Oh, she has reappeared!

わらわは、この国に『紙漉き』を伝えた「川上御前」の化身。和紙を愛する人々に時空を越えて会いにくるのじゃ。

Ⓚ I'm the incarnation of "Kawakami-Gozen", who gave the technique of "Kami-suki (paper-making)" to the Japanese people. I come to see 'Washi-lovers' throughout time and space.
⑧

そのスゴさ、よかったら見に行かんかー？

⑩ Ⓚ Shall I take you to see the great world of Washi?

なるほど！…って、全くナゾだらけだけど、なんだかスゴい世界かも？

Ⓢ Hmm, I can't understand this situation at all, but Washi seems to have a great world.
⑨

Come on,
Let's Go!
ワシが案内
するぞ。

…あれ、ワカミちゃんは？

謎に包まれた自称・和紙の神の化身と、ヘンな英国人・グランパに導かれるRPG的展開の渦中へ。ショータの行く手に待つのは、果たして…？

Ⓖ I'll guide you to the world of Washi.
Ⓢ Yeah, Let's go! But, where's Wakami?

What will happen to Shota? Led by a mysterious incarnation of the Goddess of Washi and the strange British gentleman?
⑪

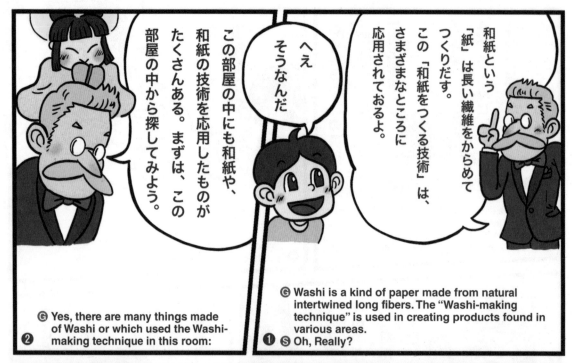

和紙という「紙」は長い繊維をからめてつくりだす。この「和紙をつくる技術」は、さまざまなところに応用されておるよ。

へえ そうなんだ

この部屋の中にも和紙や、和紙の技術を応用したものがたくさんある。まずは、この部屋の中から探してみよう。

Ⓖ Washi is a kind of paper made from natural intertwined long fibers. The "Washi-making technique" is used in creating products found in various areas.
❶ Ⓢ Oh, Really?

Ⓖ Yes, there are many things made of Washi or which used the Washi-making technique in this room:
❷

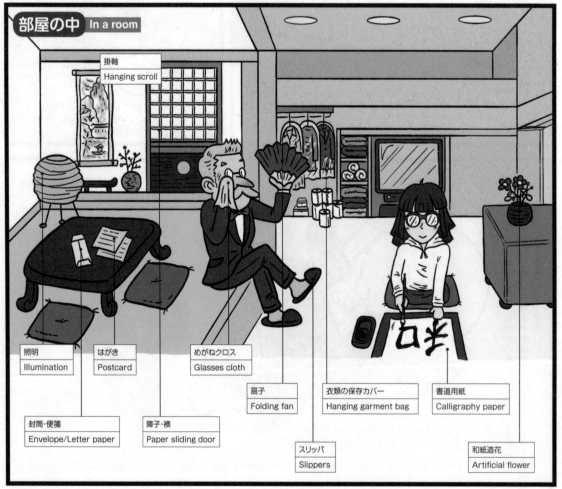

部屋の中 In a room

掛軸
Hanging scroll

照明
Illumination

はがき
Postcard

めがねクロス
Glasses cloth

扇子
Folding fan

衣類の保存カバー
Hanging garment bag

書道用紙
Calligraphy paper

封筒・便箋
Envelope/Letter paper

障子・襖
Paper sliding door

スリッパ
Slippers

和紙造花
Artificial flower

職員室 Teacher's lounge

カレンダー
Calendar

インクジェットプリンタ用紙
Inkjet printer paper

スマホカバー	名刺	俳句帳	ブックカバー	ノート	筆箱
Smartphone cover	Business card	Haiku book	Book protector	Notebook	Pencil case

披露宴 Wedding reception

敷き紙
Paper table mat

ドレス
Dress

名札	メニュー	箸袋	お札	靴下	帽子	芳名帳
Nametag	Menu	Chopsticks bag	Banknote	Socks	Hat	Guest book

おみやげ売り場 Souvenir shop

タペストリー
Tapestry

凧	懐紙	人形	バッグ	折り紙
Kite	Kaishi (a paper packet)	Doll	Bag	Origami

文化財の修復 Restoration of cultural assets

文化財の補修紙
Repair paper of cultural assets

Ⓢ First, what is Washi made from?
❶ Ⓖ Washi is made from fibers of tree bark. Let's have a look at these materials

楮（コウゾ） Kozo (Paper Mulberry)

コウゾは、くわ科の落葉低木。栽培が容易で毎年収穫でき、和紙に一番多く使われています。

Kozo is a deciduous shrub in the family of mulberry. It's an easy-to-grow plant and harvested yearly, so it's the most-used raw material of Washi.

三椏（ミツマタ） Mitsumata (Oriental Paperbush)

ミツマタは、じんちょうげ科の落葉低木。枝先が3つに分かれることからミツマタの名前がつきました。苗を植えてから3年ごとに収穫して使います。

Mitsumata is a deciduous shrub. The branches are split into three stems, so this plant was named Mitsumata (three-pronged tree). It takes 3 years to harvest for Washi-making.

雁皮（ガンピ） Ganpi

ガンピは、じんちょうげ科の落葉低木。ただ、成長が遅く、栽培が難しいので、野生のものを採集して使います。

Ganpi is a deciduous shrub, as Ganpi isn't easy to grow, wild plants are used for Washi-making.

黄蜀葵（トロロアオイ） Tororo-aoi (Aibika plants)

トロロアオイは、アオイ科の植物。この根から採取される粘液はネリと呼ばれ、和紙を漉くときに欠かせない植物です。

Tororo-aoi is a malvaceous plant. The mucilage extracted from its root is called Neri, which is an indispensable ingredient for Washi-making.

和紙とは、古くから日本でつくられてきた紙のことさ。

なあ、御前。

③ ⑥ Washi is traditional Japanese paper, isn't it, Gozen?

和紙と洋紙はどう違うの。

洋紙

和紙

⑤ ⑤ What's the difference between Washi and Yohshi (Western Paper)

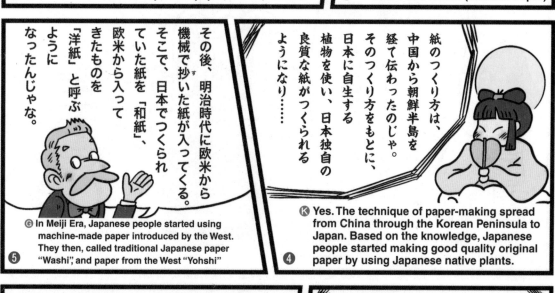

その後、明治時代に欧米から機械で抄いた紙が入ってくる。そこで、日本でつくられていた紙を「和紙」、欧米から入ってきたものを「洋紙」と呼ぶようになったんじゃな。

⑥ In Meiji Era, Japanese people started using machine-made paper introduced by the West. They then, called traditional Japanese paper "Washi", and paper from the West "Yohshi"

⑤

紙のつくり方は、中国から朝鮮半島を経て伝わったのじゃ。そのつくり方をもとに、日本に自生する植物を使い、日本独自の良質な紙がつくられるようになり……

Ⓚ Yes. The technique of paper-making spread from China through the Korean Peninsula to Japan. Based on the knowledge, Japanese people started making good quality original paper by using Japanese native plants.

④

ただ、その後、機械抄き和紙も作られるようになった。この本もそうだよ。均一で大量に必要な和紙は、機械抄きのほうが合っておるからな。

ふーん

⑥ However, we now have machine-made Washi, even this booklet! Machine mass production is suitable for Washi-making as it ensures quality and uniformity.

⑦

手漉き和紙は手作りなので、温かみや味がある。一方、洋紙のような機械抄き紙は、大量生産が可能で、安価なのが特長じゃ。

Ⓚ Hand-made Washi has a feeling of warmth, texture and style, it is tasteful in design, whereas machine-made paper like Yohshi is inexpensive and mass produced.

⑥

でも植物からできている和紙は、他の紙と違って、どこがすぐれているの。

和紙の特長を紹介しよう。全国各地でさまざまな古文書（こもんじょ）が今も残っているのは、和紙がもつすぐれた保存性のおかげじゃ。

Ⓖ Let me explain the main characteristics of Washi, Washi is long-lasting, so it preserves even old paper documents in good condition.

❶

Ⓢ But, how is Washi different from other types of paper?

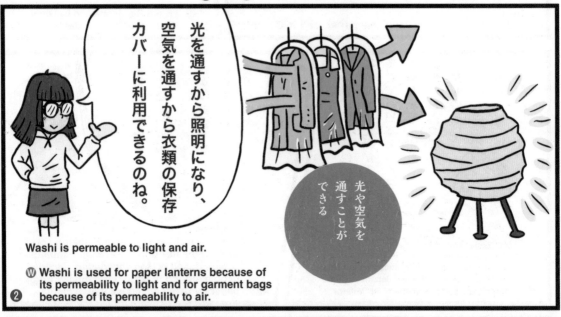

光を通すから照明になり、空気を通すから衣類の保存カバーに利用できるのね。

光や空気を通すことができる

Washi is permeable to light and air.

Ⓦ Washi is used for paper lanterns because of its permeability to light and for garment bags because of its permeability to air.

❷

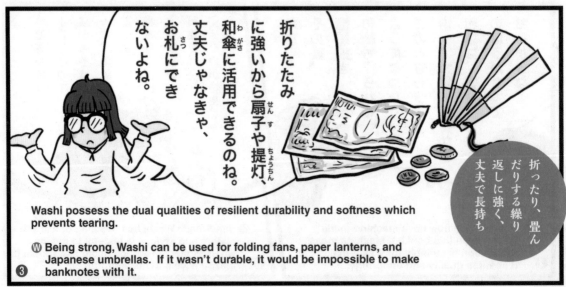

折りたたみに強いから扇子や提灯、和傘（わがさ）に活用できるのね、丈夫じゃなきゃ、お札（さつ）にできないよね。

折ったり、畳んだりする繰り返しに強く、丈夫で長持ち

Washi possess the dual qualities of resilient durability and softness which prevents tearing.

Ⓦ Being strong, Washi can be used for folding fans, paper lanterns, and Japanese umbrellas. If it wasn't durable, it would be impossible to make banknotes with it.

❸

障子に霧吹きをして、乾いたらピンと張るのは、元に戻る性質を利用しているのね。

水を含んでも元に戻る性質がある

Washi restores after drying, even if it gets wet.
Ⓦ This characteristic is one reason it is used for repapering Shoji (Japanese sliding doors).
④

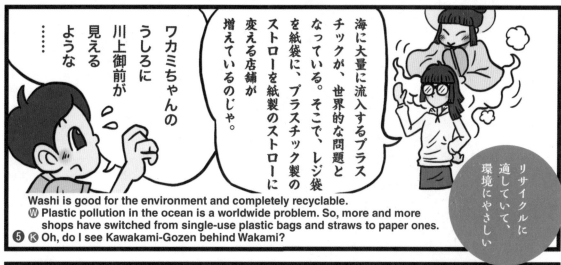

ワカミちゃんのうしろに川上御前が見えるような……

海に大量に流入するプラスチックが、世界的な問題となっている。そこで、レジ袋を紙袋に、プラスチック製のストローを紙製のストローに変える店舗が増えているのじゃ。

リサイクルに適していて、環境にやさしい

Washi is good for the environment and completely recyclable.
Ⓦ Plastic pollution in the ocean is a worldwide problem. So, more and more shops have switched from single-use plastic bags and straws to paper ones.
⑤ Ⓚ Oh, do I see Kawakami-Gozen behind Wakami?

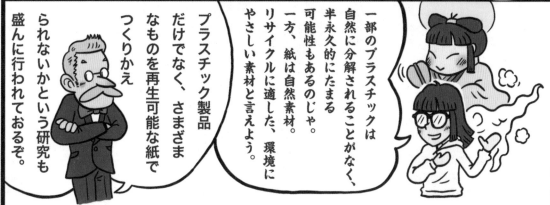

プラスチック製品だけでなく、さまざまなものを再生可能な紙でつくりかえられないかという研究も盛んに行われておるぞ。

一部のプラスチックは自然に分解されることがなく、半永久的にたまる可能性もあるのじゃ。一方、紙は自然素材。リサイクルに適した、環境にやさしい素材と言えよう。

Ⓦ Most plastic products aren't biodegradable, but Washi is made of natural materials. So, the paper is recyclable and good for the environment.
Ⓖ Now, there are many studies for making various products with paper instead of other less environmentally friendly materials.
⑥

和紙の
すばらしさが
世界に認められて、
２０１４年に、
国の重要無形文化財に
指定されている
細川紙（埼玉県）、
本美濃紙（岐阜県）、
石州半紙（島根県）の
３つが、「和紙：日本の
手漉和紙技術」として
ユネスコ無形文化遺産
に登録されたぞ！

Ⓖ Washi is acknowledged around the world as an extraordinary craft unique to Japan. Three types of Washi - "Sekishu-Banshi (In Shimane)," "Hon-minoshi (In Gifu)" and "Hosokawa-shi (In Saitama)," were registered on the representative list of UNESCO's Intangible Cultural Heritage of Humanity as "Washi, craftsmanship of traditional Japanese hand-made paper" in 2014.

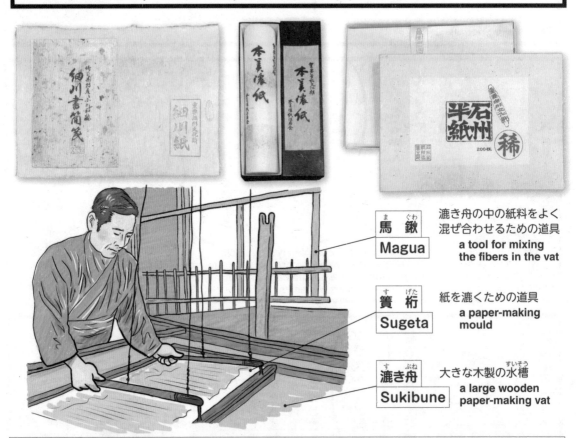

馬鍬 Magua	漉き舟の中の紙料をよく混ぜ合わせるための道具 a tool for mixing the fibers in the vat
簀桁 Sugeta	紙を漉くための道具 a paper-making mould
漉き舟 Sukibune	大きな木製の水槽 a large wooden paper-making vat

川上御前の伝説 （The legend of Kawakami-Gozen）

昔々、大昔　越前の国（今の福井県）の五箇と呼ばれる貧しい村里に一人の美しい女性が現れました。女性は村人に向かって「この地は清らかな水に恵まれているから、紙を漉いて暮らすとよいでしょう」と言って、紙漉きの技術を教えてくれました。村人は感謝して名を尋ねましたが、「岡太川の川上に住むもの」と言って消えていき、再び現れることはありませんでした。村人は女性を「川上御前」と呼び、その後岡太神社を建てて今日に至るまで「紙の神様」（紙の祖神）としてお祀りしています。

Once upon a time, a beautiful woman appeared in Goka, a poor village in Fukui. The woman said to the villagers they should make paper because there was a lot of clear water flowing near their village and so, she taught them the art of paper-making. When the villagers asked her name, the woman replied, "I'm the one who lives in the upriver region of Okamoto", she then disappeared forever. The village prospered from her guidance, in appreciation the villagers built Okamoto Shrine in order to worship the woman, whom they called "Kawakami-Gozen", the Goddess of Paper.

第2章 和紙ができるまで（楮紙の場合）

The Washi-making Process (Kozo paper)

1. 楮の刈り取り／Harvest of Kozo (raw materials)

１年間で３〜４ｍに成長した楮を毎年冬に刈り取ります。そして一定の長さに切りそろえて、束にします。

Kozo, which grows 3-4m height in a year, is harvested every winter. The branches are cut to equal lengths, gathered and bundled.

十分に水を吸わせた植物の繊維をシート状にして乾燥させると繊維同士がくっついて、紙になるんじゃ。

植物の繊維からどうやって紙ができるの？

Ⓖ Paper is made when plant fibers intertwine by flattening into sheets and drying the fibers.

Ⓦ How do we make paper from plant fibers?

2. 蒸す／Steaming branches

楮の皮をむきやすくするために、蒸します。

Branches of Kozo are steamed in order for the bark to be easily removed, as only the bark is used to make Washi.

アジアでは、中国、韓国、台湾、タイ、ブータン、ネパール、インドなどで作られ、イタリアやフランスなどのヨーロッパでも作られておるのじゃ。

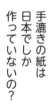

手漉きの紙は日本でしか作っていないの？

Ⓚ Not only Japan but other Asian countries like; China, Korea, Taiwan, Thailand, Bhutan, Nepal, and India. Also, European countries like Italy and France.

Ⓢ Is handmade paper made only in Japan?

3. 表面の黒皮を削る／Scraping outer black bark

はぎ取った楮の皮は、包丁をあてて表面の黒皮を削り取ります。
残りの部分が和紙の原料になります。

The bark of Kozo has several layers. The outer layer of black bark (Kurokawa) is carefully scraped off with a knife. The remaining part of bark is a raw material used for the Washi-making process.

世界各国で手漉きの紙が作られているのね。

Ⓦ Handmade paper is fabricated in many countries!

中国では青檀、稲藁、竹など、タイではカジノキなど、インド、ヨーロッパではコットンなどが使われておるんじゃ。

Ⓖ Paper is also made from Qing tang (winged celtis), straw and bamboo in China, from Kajinoki (paper mulberry) in Thailand and from cotton in Europe.

4. 楮の皮を干す／**Drying bark**

黒皮を削り取った残りの皮は、乾燥させる
と保存できます。

Bundles of bark are hung to dry.
After drying, they are baled and stored.

だから
さまざまに
異なった
風合いの紙が
できるんだね。

手漉きの紙には、
原料の違いからも
各国それぞれ
特徴があるとも
いえるのじゃ。

Ⓢ So, there are many kinds of paper which have
different textures.

Ⓚ There are many kinds of hand-made paper,
which have unique characteristics depending
on the different raw materials used.

5. 川に晒す／Soaking in the river

皆を流水につけ、天然漂白します。
紙を作るには、たくさんのきれいな水が必要です。

Dried bark is soaked in the stream for natural bleaching from exposure to water and sunlight. Washi-making requires a large amount of clear water.

川の水や雪に晒して日光にあてる。冷水と日光が紙を白くすると言われとるぞ。今はほとんど川に晒さず水槽などを使っているんじゃ。

Ⓖ The tree bark is soaked in river water or placed on the snow to be exposed to sunlight. It is said that cold water and sunlight make the paper white. Nowadays, the bark is soaked in a water tank instead of the river.

楮を原料とする和紙は、自然な白さと丈夫な紙質が特徴といわれているのよね。

Ⓦ Washi made from Kozo has characteristics of natural whiteness and durable paper quality.

6. 皮を煮る／**Boiling the bark**

石灰水・木灰汁などで煮て、楮の皮をやわらかくします。

The bark is boiled in a large caldron filled with a limewater or lye until it becomes soft.

一人前の和紙職人になるには10年以上の修行が必要と言われるが、年々、和紙職人の数が減少しておるのじゃ。

和紙作りはたいへんそうだね。

K It requires more than 10 years of practice to become a skilled Washi craftsman, and the number of craftsmen decreases year after year.

S Washi-making looks like hard work.

7. 灰汁抜き、ちり取り／**Removal of impurities**

煮た皮を流水につけます。
水の中で、ちりを手で丹念に取り除きます。
手間のかかる作業です。

The bark is soaked in a cold stream in order to be rinsed by the flow of water. During this process, any remaining parts of black bark and dust are carefully removed by hand. It is certainly the most time-consuming part of the process.

日本人はそうやって紙つくりを大切に続けてきた。昔から和紙の産地では和紙に関する祭りも行われている所もあるんじゃ。

Ⓖ Japanese people have been patiently making Washi like that for centuries. Some Washi production areas even have festivals related to Washi.

原料を水に晒して煮て灰汁抜き、ちり取り……。たいへんな作業ね。

Ⓦ Soaking in water, then boiling, removing scum and dust…what hard work!

8. 楮を打つ（叩解）／ **Pounding**

皮の繊維を棒で叩いて、繊維をほぐします。
ようやく和紙の原料ができあがりました。

The bark is pounded with a hard-wooden bat, this process allows the fibers of bark to become loose and soft. It takes a lot of work and patience to produce the materials needed to create Washi.

紙漉き体験が
できる施設は
全国にたくさん
あるぞよ。
調べてみると
よい。

僕も
和紙作りを
体験して
みたいな。

Ⓚ There are many places you can go to experience and practice Washi-making. Look for where you can find them.

Ⓢ I wanna try making Washi.

9. トロロ叩き ／ Tororo beating (Neri making)

トロロアオイ（黄蜀葵）の植物の根を叩き、
粘液（ねんえき）を採ります。これをネリといいます。

Washi-making requires the use of mucilage called Neri, which is principally extracted from the roots of Tororo-aoi plants (Aibika plants).

ネリ（粘液）が
繊維を水中で均一に
分散させ沈殿をおさえる。
それを簀桁（すげた）の上でゆすって
繊維同士を絡（から）めることで、
薄くても強度のある紙が
できるんじゃ。

植物からとった
繊維を
水の中に入れて、
「ネリ」と呼ばれる
粘り気のある液体を
混ぜるのね！

Ⓖ Neri (mucilage) causes the fibers to evenly disperse in water. The thin but strong paper can be made by moving Sugeta with this liquid back and forth to entangle the fibers.

Ⓦ The next process is putting plant fibers in water and mixing them with "Neri" mucilaginous liquid.

10. 紙漉き／Sheet formation

漉き舟の中に叩解した繊維・水・ネリを入れ、かき回し紙料をつくり、それを簀桁ですくい上げます。漉き方には、「流し漉き」と「溜め漉き」があります。

The pounded fibers are mixed with Neri and water in a vat called Sukibune. Then, this mixture is scooped onto the Sugeta (paper-making mould) in order to form a sheet of paper. There are two types of sheet forming methods: Nagashi-zuki, which is mainly used for making thin paper, and Tame-zuki, which is mainly used for making thick paper.

流し漉きは原料を数回簀桁にすくい取り前後左右にゆすって作るやり方。溜め漉きは、原料を簀桁の上にすくい取り水が抜けるまで待つやり方じゃ。

Ⓚ Nagashi-zuki, is a method of repeatedly scooping a little amount of fibers and moving the mould back and forth, and Tame-zuki is a method of scooping fibers one at a time and waiting until the water has dripped through the screen.

流し漉きと溜め漉きはどう違うの？

Ⓢ What's the difference between Nagashi-zuki and Tame-zuki?

11. 紙を絞る ／ Pressing the paper

漉いた紙を積み重ねたものを紙床といいます。紙床に積まれた紙の水分を一晩かけてゆっくり絞っていきます。

The newly formed sheets of wet paper are piled up. The stack of wet sheets is called Shito (paper beds) and are slowly compressed overnight for dehydrating.

トロロアオイで作ったネリ（粘液）の働きで紙が一枚一枚、はがれるのじゃ。

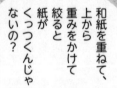

和紙を重ねて、上から重みをかけて絞ると紙がくっつくんじゃないの？

Ⓚ No, the effect of Neri (mucilage) enables each sheet of Washi not to stick together.

Ⓢ When the sheets of Washi are piled and dehydrated by weighing down, they stick together, don't they?

12. 乾燥／**Paper drying**

紙床から紙を一枚ずつはがし、刷毛で板にはって天日で乾かします。これを天日干しといいます。天日で自然乾燥した紙はしなやかで独特な風合いがあります。

The pressed paper sheets are removed one by one from the stack and stretched out on wooden boards with a special brush to be dried in the sun. This process is called Tenpi-boshi (sun drying). The sundried paper becomes soft, yet, resilient and has a unique texture.

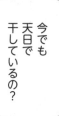

今では屋内で乾燥機を使って乾かす方法もある。雨の日でもたくさんの紙を乾かせるんじゃ。

G Now, there is also a method of using a dryer. This method allows us to dry many sheets of Washi even on rainy days.

今でも天日で干しているの？

W Is Washi dried in the sun still today?

13. 検品(けんぴん)／Quality checking

乾燥した紙は、一枚一枚検査をします。破れた不良紙などは漉き返され再利用します。

After drying, each sheet of paper is carefully checked based on its thickness, color, and texture. Any imperfections, such as: breaks, shrinkage, rough surfaces or uneven thickness etc. are rejected, re-boiled and recycled under the same process of Washi production.

簀(す)に模様をぬいつけて紙を漉(す)くと模様の部分だけうすくなり、光にかざすと白く透けて見えるんじゃ。

Ⓖ Watermarked paper is made by sewing marks on a screen and getting the mark areas thinner.

透(す)かしはどうしたらできるの？

Ⓦ Please tell me how to make watermarked paper.

14. 紙を裁つ、仕上げ／Cutting and packaging

検品が終わった紙は枚数をそろえて積み重ね、あて板（定規）をあてて定められた寸法に裁断します。
そして、梱包して商品の完成です。

Quality checked Washi is then cut to size and packaged. This is the final process of Washi-making.

紙は、
糸へんに氏。
植物繊維（糸）を
たいらで
なめらかにする（氏）
という意味なのじゃ。

なぜ
紙という字が
できたの？

Ⓚ The Kanji 紙 (paper) consists of 2 parts: 糸 (fibers) and 氏 (flat and smooth). So, this Kanji means a substance of flat and smooth fibers.

Ⓢ Can you explain the meaning of the Kanji character for paper?

第２章「紙ができるまで」は『紙漉重宝記』を参考にして描かれたものです。

『紙漉重宝記』とは１７９８（寛政10）年に、石見国（いまの島根県）の国東治兵衛が著した製紙の解説書です。原料の刈取りから製品の出荷に至るまでを詳細に図解した製法書で、当時の画家靖中庵丹羽桃渓の挿絵が多数入っています。

We described the Washi-making process of Chapter 2 with reference to "Kamisuki Choho-ki", a "Guidebook of Paper-Making" written by Jihe Kunisaki of Shimane in 1798. This book describes the Washi-making process in detail with many illustrations of Tokei Niwa.

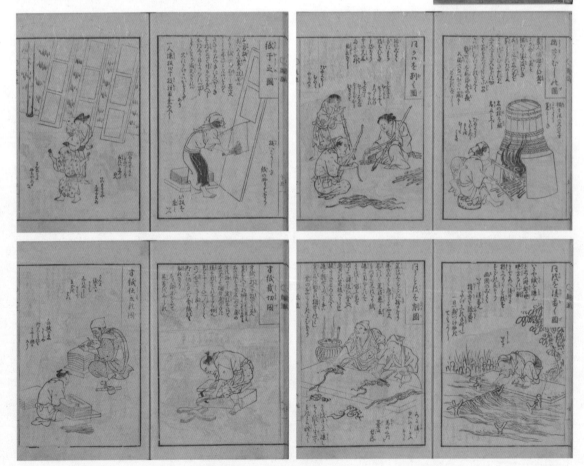

第3章

和紙と暮らしの歴史

Japanese History and Washi

画：岡田 潤

1. 奈良時代／Nara Period

　紙は中国で発明されました。日本へ紙のつくり方が伝わったのが610年と「日本書紀」に記載されています。日本に残る最古の紙（702年）は正倉院に保管されています。

　当時の紙の使用目的は、戸籍作成などの公文書と写経でした。経文奉納で有名なのは「百万塔」です。

　木製の塔（約15cm）は百万基つくられ、なかには印刷された経文「陀羅尼経」が1枚ずつ納められていました。これは日本最古（770年）の印刷物とされています。

Paper was originally invented in China.
It was in 610 AD that the paper-making technique was introduced to Japan, according to Nihon Shoki, the ancient Chronicles of Japan. The oldest existing paper in Japan is preserved in the Shosoin Repository in Nara. Paper was mainly used for official documents like family register and for Buddhist sutras. One of the well-known votive sutras is "Hyakumanto (One million pagodas)". It consists of one million 15cm tall wooden pagodas in which each printed sutra called "Darani-kyo" is stored individually. It is said to be the oldest printings in Japan (770 AD).

2. 平安時代／**Heian Period**

平安時代（794～1192）になると、貴族の間で和歌や漢文・書・絵巻が盛んになり、趣のある紙が求められました。

紫式部や清少納言など女流作家たちが誕生し、宮廷の雅な様子が美しい仮名文字でつづられました。

随筆や物語では、「枕草子」「源氏物語」など、絵巻では、「源氏物語絵巻」「信貴山縁起絵巻」などがあります。

また、平安時代を代表する美しく加工された料紙で作られた歌集「西本願寺三十六人家集」も著されます。

ただ、紙は高級品で、一般の人には手の届かないものでした。

In the Heian Period (794-1192) aristocrats loved classical Japanese poetry (Waka), Chinese poetry, calligraphy and picture scrolls, in turn, they had a great desire for tasteful paper. There were also female authors who described an elegant atmosphere of the Imperial Court in Japanese Kana characters on paper: Murasaki Shikibu wrote "The Tale of Genji", one of the most famous Japanese romance, and Sei Shonagon wrote "The Pillow Book", one of the most famous Japanese informal essays. Picture scrolls like "Genji Monogatari Emaki" and "Shigisan Engi Emaki" were also created.

"Nishi Honganji Sanju-rokunin Kashu" was written on Ryoshi, one of the most beautifully processed papers for calligraphy. However, at that time paper was such a luxury that common people could not obtain it for personal use.

3. 鎌倉時代／**Kamakura Period**

かまくらじだい

　鎌倉時代（1192〜1333）に入ると、貴族に変わり武士が権力を持つようになり、特に経文などの木版画印刷が盛んに生み出されました。
　武士の間では紙は実用的に使用されました。書状も数多く残されています。
　また、似絵という写実を重んじる絵画が生まれ、武将、高僧の似顔絵も描かれました。

In the Kamakura Period (1192-1333), the Samurai warriors took power from the aristocrats. Many wooden block prints of Buddhist scriptures were produced. Among Samurai, paper was mainly used for practical purposes, there remains a lot of letters written from that time.
Nise-e (likeness picture) introduced the realistic depiction, paintings portraying Busho, Samurai commanders, and high-ranking Buddhist priests.

4. 室町時代／Muromachi Period

　室町時代（1336～1573）になると、銀閣寺に代表される建築様式に、書院造りが誕生します。
　建具として襖、障子のほか、屏風や衝立などに紙が用いられました。山水画や仏教画なども盛んになり、掛け軸、屏風絵、巻物などの水墨画が残されました。

In the Muromachi period (1336-1573) a new architectural style of Samurai culture called Shoin-zukuri represented by Ginkakuji Temple was born. Paper was now being used for sliding doors, folding screens, partitions etc. For decorative purposes landscape pictures and Buddhist impressions were painted on screens and hanging picture scrolls.

5. 安土桃山時代／Azuchi-Momoyama Period

安土桃山時代（1573～1603）は、戦乱の終結と天下統一の気運の中、南蛮屏風に見られるように豪壮・華麗な文化が花開きました。

その一方で、簡素で質素な「わび」を大事にした茶の湯の文化も生まれます。

このように、上代から盛んに描かれた仏画や絵巻物は、社会構造や生活様式の変化にともない、多彩な発展を遂げていき、こうした美術の隆盛は、製紙技術の向上にも大きな影響を与えることになりました。

そして、海外との貿易も本格的に実施され、朱印状をもつ朱印船によって、日本の紙は海外に渡っていきました。

In the Azuchi-Momoyama period (1573-1603) at the end of warfare and increasing momentum of the unification of Japan, luxurious and magnificent cultural representations of exotic folding screens flourished. In contrast, derived a culture of 'tea ceremony' in pursuit of simplicity and silence called "Wabi" was also born. The Buddhist pictures and illustrated handscrolls, created from ancient times emerged in various ways with the change in social structures and lifestyle. The prosperity of fine art influenced the development of the paper-making technique. Full-scale overseas trade began, and Japanese paper started to spread around the world by the exportation of Shuinsen, merchant ships with trading patent rights granted by the government.

6. 江戸時代Ⅰ / Edo Period

江戸時代（1603〜1868）を通じて、各藩で製紙を奨励するようになり、専売制とする藩も現れました。

そして紙の生産が盛んになるにつれ、紙は庶民の暮らしに浸透し、江戸文化が花開いていきました。

絵にあるのは、藩札・帳面類・提灯・薬袋・瓦版などです。

火事の多かった江戸の町では、火がでると大福帳などを水の中に投げ入れました。和紙は、水に強いので、後で乾かすと墨で書いてある文字や数字は読めるのです。火消しが着る半纏は、紙糸を織り、漆を塗って使うこともありました。

Throughout the Edo Period (1603-1868), each feudal domain encouraged paper manufacturing, some monopolized the making and selling of paper. As the manufacturing of paper evolved, paper became more common and familiar to people in everyday life, this 'paper culture' supported by townspeople flourished. In this picture, there are local banknotes, notebooks, paper lanterns, medicine packages and Kawara-ban (newspaper). In the city of Edo, where fires regularly occurred people threw their account books into water when fires broke out. As Washi is water resistant, even when wet, letters and numbers written in Japanese ink are readable after the books dried. Hanten, firefighters' uniforms were sometimes made of Washi fabric finished with Urushi (Japanese lacquer).

7. 江戸時代Ⅱ／Edo Period (cont'd)

えど じだい

　政治が安定し、町人文化が栄えた江戸時代は、生活のすみずみにまで紙が行き渡りました。子どもたちは寺子屋で手習いをし、また、人形・千代紙・風船・凧・カルタなど紙で作られた玩具で遊びました。

　江戸文化の隆盛は、多くの書物を生み、浮世絵版画は、紙の加工技術を発展させました。絵の中でたすきをして刷毛を持っている人は、表具師です。絵や書を額に入れて飾る作業をしています。

As the government stabilized and urban culture supported by townspeople flourished, paper became increasingly widespread throughout everyday life. Children learned calligraphy in Terakoya (private temple schools) and played with paper toys such as dolls; paper with colored figures called 'chiyogami', paper balloons, kites and playing cards. The prosperity of this culture also gave rise to a lot of books and documents. Ukiyo-e, Japanese woodblock printings also promoted the development of paper processing techniques. The man using a wide brush in this picture is a Hyogu-shi, whose job it is to place paintings and calligraphy works on frames or scrolls for display.

8. 江戸時代Ⅲ／Edo Period (cont'd)

え　ど　じ　だい

　江戸時代、紙は人々の生活を豊かにし、いろいろな場所で活躍していきます。雨にも強く、傘や雨合羽に使われるほか、寺や神社の護符や四季折々の行事など、家の外でも、需要は高まります。

　強さと美しさを兼ね備えた紙は、絵にある七夕飾りの他、お面・うちわ・提灯などと、用途を広げていきました。

Paper enriched everyday life of common people and played an important role in many areas of daily activities. Being water resistant, paper was used for umbrellas and raincoats, in addition, demand for using paper increased even outside the home. Paper was not only used for talismans of temples and shrines but also for tools in many seasonal events. As Washi had both strength and beauty, its application expanded for use in masks, fans, paper lanterns, ornaments and decorations for Tanabata festival (celebrating stars and summer).

9. 明治時代／Meiji Era

めい じ じ だい

明治（1868 〜 1912）に入ると、西洋の技術が導入され、洋紙が作られるようになりました。文明開化の時代を迎え、新聞・雑誌など報道の分野が急速に発展し、新しい紙の需要が生み出されました。

明治18年（1885年）には、日本銀行の紙幣が印刷されます。

With the arrival of the Meiji Era (1868-1912), the western technique for paper-making was introduced and machine-made paper came into production. This began the age of westernization and modernization in Japan, press reports such as newspapers and magazines spread rapidly and demand for new kinds of paper were created. In turn, in 1885, the Bank of Japan started printing its first paper banknotes.

10. 大正時代／Taisho Era
（たいしょうじだい）

　大正時代（1912～1926）には、生活様式が洋風化されていきました。西洋文化と日本の文化がまじりあい、大衆向けの新聞・文芸誌・娯楽誌が普及し、「大正ロマン」といわれる文化が生まれました。

　また、子どものための読み物・冊子も相次ぎ創刊されました。

In the Taisho Era (1912-1926), lifestyles throughout Japan had been westernized, as Japanese and western cultures mixed, newspapers, literature and magazines for the masses were popularized and the "Taisho Roman" culture flourished. In addition, books and booklets for children were published one after another.

11. 昭和時代Ⅰ／Showa Era
しょうわじだい

　昭和（1926～1989）の前半、紙は戦時体制で、統制経済の下におかれました。戦後十年ほどたつと日本経済は立ち直り、新しい時代を迎えます。

　テレビ・映画など大衆文化とスポーツの発展にともない、出版物が増え、印刷技術の発展・高度化により、紙、特に洋紙の需要がますます増えていきました。

　また、教科書・書籍・ノートなどの教育文化、子どもたちの遊びの道具やお菓子の箱など、生活用品に多種多様な製品が生み出されました。

In the first half of the Showa Era (1926-1989), paper was economically controlled on a war footing. Within 10 years after the war, the Japanese economy recovered, and more prosperous days returned. Public cultural interests including TV programs, films, sports publications, books, magazines and comics popularized. The development and sophistication of printing techniques enlarged the demand for machine-made paper. With this demand, a variety of products were created for school supplies; textbooks, books and notebooks and household goods; toys and sweets boxes for children.

12. 昭和時代Ⅱ／*Showa Era (cont'd)*

しょう わ じ だい

昭和も40年代（1965年〜）になると、宣伝・広告・
出版・新聞など、印刷物はいっそう増え、日常生活の
身の回りの紙には洋紙が多く使われました。

一方、和紙の伝統を守り継ごうという機運が大きく
高まり、政府や自治体は、手漉き和紙の一部に重要文
化財・文化財保存技術の指定を行うようになりました。

また、世界でも、美術館・博物館などの資料の保
存・修復に和紙が使われるようになり、日本の和紙は
「WASHI」として、世界中に知られるようになりまし
た。

From 1965 onwards, publicity, advertisements and
newspaper publications became more popular among
people and machine-made paper was widely used
in daily life. At the same time, the momentum for
maintaining and carrying on the traditional aspects of
Washi highly increased. The Japanese government
decreed the art of traditional 'handmade' Washi as
an Important Cultural Properties or an Conservation
Technique for Cultural Properties Japanese traditional
paper played an integral part in the conservation and
restoration of cultural assets in museums around the
world, becoming widely recognized and known as
"WASHI" globally.

13. 平成時代Ⅰ／ Heisei Era

　平成（1989〜2019）に入ると、企業ではコピー用紙やパソコンなどの記録紙がさらに大量に使用され、その多くに再生紙が使われています。品物の運搬に使用される段ボール箱なども同様です。

　紙のリサイクルは、再資源化・省エネルギー、さらには環境保護につながっています。

In the Heisei Era (1989-2019), paper for copying and electronic documents were used in large amounts, most deriving from recycled paper like cardboard boxes. Paper recycling ultimately leads to energy saving and protection of the environment.

14. 平成時代 Ⅱ ／ Heisei Era (cont'd)

　2014年（平成26年）国連教育科学文化機関（ユネスコ）は、「和紙：日本の手漉和紙技術」を無形文化遺産に登録しました。

　和紙は、書道・絵画・美術工芸・保存修復用の他、その機能を活かしたインテリアなどになくてはならない存在として、日本はもちろん世界で高い評価を得ています。

　現在は、デジタルメディアが記録媒体の主流です。しかし、情報機器の耐久年数はその寿命と合わせて不明確で一瞬にして消失する危険性をはらんでいます。後世に情報・記録を伝達するには紙、特に保存性のすぐれた和紙こそが最適と思われます。

2014, UNESCO registered "Washi, craftsmanship of traditional Japanese hand-made paper" as an Intangible Cultural Heritage. Washi is highly valued not only in Japan, but also, throughout the world as one of the necessary elements for Shodo (Japanese calligraphy), painting, fine arts and crafts, conservation and interior restoration, etc. Nowadays, digital media is the mainstream for recording data, however, electronic or digital storage cannot be guaranteed over a long period of time and has a risk for sudden loss of data. With all this advancement in technology, it is paper, in particular Washi which has stood the test of time, remaining the most appropriate medium to pass our information and records down through the generations.

15. 令和時代／Reiwa Era

手漉き和紙の驚くべきところは、過去からずっと基本的に製法が変わらないことです。

それにもかかわらず、和紙の製法を活かした製品は、常に時代に即した新しさを提供しています。

時代が令和（2019〜）になっても和紙は、2020年東京オリンピックの表彰状に取り上げられるなど、ますます注目されています。

最近の主な新聞記事から、伝統を備え革新に満ちた和紙の注目ぶりを紹介します。

Surprisingly, even in the Reiwa Era (2019~), the methods of making hand-made Washi remain nearly unchanged from ancient time.
Nevertheless, products using the Washi-making methods are always in response to the needs and trends of the times.
Washi attracts more and more attention from the outside world. For example, the testimonials made of Washi have been used for the Tokyo Olympic Games in 2020.

岐阜新聞2019年7月25日付掲載
讀賣新聞2019年8月 7日付掲載
讀賣新聞2019年8月20日付掲載

第4章
広がる和紙の世界

More information about Washi

和紙の産地

和紙の産地は日本全国に

北海道
① 笹貴紙

岩手県
② 富貴紙
③ 東山和紙

山形県
④ 成島和紙
⑤ 月山和紙
⑥ 長谷川和紙工房
⑦ 深山和紙

宮城県
⑧ 白石和紙

福島県
⑨ 遠野和紙

東京都
⑩ 軍道紙(軍道紙保存会)
⑪ 東京手漉き和紙工房

埼玉県
⑫ 小川和紙(埼玉県小川和紙工業協同組合)

栃木県
⑬ 烏山和紙(合名会社福田製紙所)

茨城県
⑭ 西ノ内和紙

群馬県
⑮ 桐生和紙

山梨県
⑯ 西嶋和紙

静岡県
⑰ 駿河柚野紙(伊豆面工房)
⑱ 天城民芸湯ヶ島和紙(内藤恒雄手すき和紙記念館)

新潟県
⑲ 小国和紙(有限会社小国和紙生産組合)
⑳ 門出和紙(越後門出和紙)

富山県
㉑ 越中和紙(富山県和紙協同組合)

石川県
㉒ 加賀和紙
㉓ 能登仁行紙

福井県
㉔ 越前和紙(福井県和紙工業協同組合)

長野県
㉕ 若狭和紙
㉖ 内山和紙(内山紙協同組合)

岐阜県
㉗ 美濃和紙(美濃手すき和紙協同組合)
㉘ 美濃和紙

愛知県
㉙ 小原和紙(和紙のふるさと運営協議会)

京都府
㉚ 黒谷和紙(黒谷和紙協同組合)
㉛ 丹後和紙(田中製紙工業所)

滋賀県
㉜ 近江なる子和紙

兵庫県
㉝ 杉原紙(杉原紙研究所)
㉞ 名塩和紙(名塩手漉和紙工房)
㉟ 名塩和紙(名塩和紙工房八木)
㊱ 名塩紙(馬場製紙所)
㊲ ちくさ雁皮紙(谷徳製紙所)
㊳ 淡路津名紙
㊴ 播州ちくさ手漉和紙工房

奈良県
㊵ 吉野紙(福西和紙本舗・福西弘行)

三重県
㊶ 伊勢和紙(大豊和紙工業株式会社)

和歌山県
㊷ 山路紙(奥野誠紙漉き工房)
㊸ 高野紙
㊹ 保田紙(植和紙工房・上窪良二)

鳥取県
㊺ 因州和紙(鳥取県因州和紙協同組合)
㊻ 横野和紙(上田手漉和紙工場)

岡山県
㊼ 備中和紙

島根県
㊼ 樫西和紙
㊽ 出雲和紙(出雲民芸紙工房)
㊾ 石州半紙・石州和紙(石州和紙協同組合)
㊿ 斐伊川和紙

山口県
51 徳地和紙

愛媛県
52 伊予和紙
53 周桑和紙(東予手すき和紙振興会)
54 大洲和紙(大洲手漉き和紙協同組合)

高知県
55 土佐和紙(高知県手すき和紙協同組合)

徳島県
56 阿波和紙(阿波手漉和紙商工業協同組合)

福岡県
57 八女和紙(八女手すき和紙組合)
58 筑前秋月和紙

佐賀県
59 名尾和紙
60 重橋和紙(肥前名尾和紙)

大分県
61 竹田和紙

宮崎県
62 美々津和紙

熊本県
63 水俣和紙
64 鶴田和紙(株式会社水俣 浮浪雲工房)

鹿児島県
65 さつま和紙(さつま和紙 小迫田昭人)
66 蒲生和紙

沖縄県
67 琉球紙

出典(黒字):『和紙の手帖(2014年7月改定版7刷)』(全国手すき和紙連合会刊)
出典(赤字):全国手すき和紙連合会役員・会員名簿(2019年12月現在)
なお、ここに紹介されていない紙漉場もたくさんあります。

43 Yasuda-gami	52 Iyo Washi	
44 Inshu Washi	53 Shusou Washi	
45 Yokono Washi	54 Oozu Washi	61 Takeda Washi
46 Bitchu Washi	55 Tosa Washi	62 Mimitsu Washi
47 Kashinishi Washi	56 Awa Washi	63 Minamata Washi
48 Izumo Washi	57 Yame Washi	64 Tsuruta Washi
49 Sekishu-banshi	58 Chikuzen Akizuki Washi	65 Satsuma Washi
50 Hiikawa Washi	59 Nao Washi	66 Kamo Washi
51 Tokuji Washi	60 Jubashi Washi	67 Ryukyu-shi

Reference (Black): All Japan Handmade Papermaker's Association, "Journal of Washi", 7th Edition, July 2014
Reference (Red): Name list of All Japan Handmade Papermaker's Association (December 2019)
There are many other paper-making places in Japan.

Production area of Washi

You can find many areas of Washi throughout Japan.

① Sasa-gami	⑮ Kiryu Washi	㉙ Obara Washi
② Fuki-gami	⑯ Nishijima Washi	㉚ Kurotani Washi
③ Tozan Washi	⑰ Surugayuno gami	㉛ Tango Washi
④ Nnarushima Washi	⑱ Yugashima Washi	㉜ Ohmi Naruko Washi
⑤ Gassan Washi	⑲ Oguni Washi	㉝ Sugihara-gami
⑥ Hasegawa Washi Kobo	⑳ Kadoide Washi	㉞ Najio Washi
⑦ Miyama Washi	㉑ Etchu Washi	㉟ Najio Washi
⑧ Shiroishi Washi	㉒ Kaga Washi	㊱ Najio-gami
⑨ Tohno Washi	㉓ Noto nigyou Washi	㊲ Chikusa Gampi-shi
⑩ Gundou-shi	㉔ Echizen Washi	㊳ Awaji Tsuna-gami
⑪ Tokyo Tesukiwashi Kobo	㉕ Wakasa Washi	㊴ Yoshino-gami
⑫ Ogawa Washi	㉖ Uchiyama Washi	㊵ Ise Washi
⑬ Karasuyama Wsahi	㉗ Mino Washi	㊶ Sanji-gami
⑭ Nishinouchi Washi	㉘ Sanchu Washi	㊷ Koya-gami

和紙の資料館・美術館

和紙に限らず、紙のことがすべてわかる「紙の博物館」

公益財団法人紙の博物館は、1950（昭和二十五）年、「洋紙発祥の地」として知られる東京都北区王子に設立されました。

日本の伝統的な「和紙」、近代日本の経済発展を支えた「洋紙」を含め、国内外の紙の歴史・文化・産業を総合的に紹介しています。

正面玄関を入ってすぐのところに六畳敷きの和紙の大作「聖徳太子御影」（奥田元宋画伯）が迎えてくれます。

この見事な作品をはじめ、当博物館には、4万点の資料と1万5000点の図書を保管して展示公開する、世界でも数少ない紙専門の総合博物館なのです。

紙の歴史は古く、人類の歩みに紙が果たしてきた役割は極めて大きいといえます。

その「紙に」焦点をあて、歴史をたどり、現在を知り、未来を考える、貴重な博物館です。

紙の博物館では紙に関する資料の収集、保存、調査、研究を行うとともに、その成果を展示公開し、さまざまな教育普及事業を行っています。

所在地
東京都北区王子1の1の3　飛鳥山公園内

☎03・3916・2320

https://papermuseum.jp/ja/

紙の博物館は2020年6月に70周年を迎えるにあたりリニューアル工事で休館中です。
2020年3月17日（火）リニューアルオープン致します。

Museums of Washi

In "The Paper Museum" you will be able to learn all about the world of paper including Washi.

The Paper Museum introduces the history, culture, and industry of paper from two perspectives: Washi which is Japanese traditional paper and Western-style paper which contributed to the economic growth of modern Japan. Looking after and displaying a collection of 40,000 items and 15,000 books, we are one of the world's few comprehensive museums specializing in paper. The museum is sponsored by about 160 corporate supporters – mainly paper companies.

Asukayama Park, 1-1-3, Oji, Kita-ku, Tokyo 114-0002
TEL +81-3-3916-2320 / FAX +81-3-5907-7511
Opening Hours : 10:00 ～ 17:00(last admission at 16:30)
Closed : Mondays (except national holiday), Weekday immediately following a national holiday, New Year's holiday, period Special closure.

その他の和紙の資料館・美術館

- 紙のさと和紙資料館　茨城県常陸大宮市舟生90　☎0295・57・2252
- 烏山和紙会館　栃木県那須烏山市中央2の6の8　☎0287・82・2100
- お札と切手の博物館　独立行政法人国立印刷局　東京都北区王子1の6の1　☎03・5390・5194
- 小津史料館　東京都中央区日本橋本町3の6の2　☎03・3662・1184
- 伝統工芸青山スクエア　東京都港区赤坂8の1の22赤坂王子ビル1階　☎03・5785・1301
- 東京農工大学科学博物館　東京都小金井市中町2の24の16　☎042・388・7163
- 埼玉伝統工芸会館　埼玉県比企郡小川町大字小川1220　☎0493・72・1220
- 東秩父村和紙の里　埼玉県秩父郡東秩父村御堂441　☎0493・82・1468
- 飯山市伝統産業会館　長野県飯山市飯山1436の1　☎0269・62・4019
- 桂樹舎和紙文庫　富山県富山市八尾町鏡町668　☎0764・55・1184
- 和紙のふるさと「工芸館」　愛知県豊田市永太郎町洞216の1　☎0565・65・2151
- 小国町紙の美術博物館　新潟県長岡市小国町小国沢2531　☎0258・95・3161
- 紙の文化博物館　福井県越前市新在家町11の12　☎0778・42・0016
- パピルス館　福井県越前市新在家町8の44　☎0778・42・1363
- 卯立の工芸館　福井県越前市新在家9の21の2　☎0778・43・7800
- 富士市立博物館　静岡県富士市伝法66の2　☎0545・21・3380
- 静岡市立芹沢銈介美術館　静岡県静岡市駿河区登呂5の10の5　☎054・282・5522
- 伊豆市資料館　静岡県伊豆市上白岩425の1　☎0558・83・1859
- 山根和紙資料館　鳥取県鳥取市青谷町山根128　☎0857・86・0011
- かみんぐさじ　鳥取県鳥取市佐治町福園146の4　☎0858・89・1816
- (財)安部栄四郎記念館　島根県松江市八雲町東岩坂1754　☎0852・54・1745
- 石州館和紙会館　島根県浜田市三隅町古市場589　☎0855・32・4170
- 美濃和紙の里会館　岐阜県美濃市蕨生1851の3　☎0575・34・8111
- 伊勢型紙資料館　三重県鈴鹿市白子本町21の30　☎059・368・0240
- 黒谷和紙会館　京都府綾部市黒谷町東谷3　☎0773・44・0213
- 西宮市立郷土資料館分館　名塩和紙学習館　兵庫県西宮市名塩2の10の8　☎0797・61・0880
- 杉原紙研究所　兵庫県多可郡多可町加美区鳥羽768の46　☎0795・36・0080
- 大洲和紙会館　愛媛県喜多郡内子町平岡甲1240の1　☎0893・44・2002
- 五十崎凧博物館　愛媛県喜多郡内子町五十崎甲1437　☎0893・44・5200
- いの町紙の博物館　高知県吾川郡いの町幸町110の1　☎088・893・0886
- 八女手すき和紙資料館　福岡県八女市本町2の123の2　☎0943・22・3131
- 未来の森ミュージアム　八代市立博物館　熊本県八代市西松江城町12の35　☎0965・34・5555
- 阿波和紙伝統産業会館　徳島県吉野川市山川町川東141　☎0883・42・6120
- 紙のまち資料館　愛媛県四国中央市川之江町4069の1　☎0896・28・6257

参考：『産地別すぐわかる和紙の見分け方』久米康生著（東京美術刊）2019年11月現在／ご訪問の際は事前にお問い合わせください。

謝辞

本書の作成に当たり、ご協力およびご助言をいただきました
公益財団法人 紙の博物館 専務理事・館長の東 剛様および同
館職員の皆様、そして全国手すき和紙連合会とお世話になっ
た皆様に感謝申し上げます。

小津和紙編纂室

※本文の内容に関するお問い合わせは、
　小津和紙編纂室（電話 03-3662-1184）までお願いいたします。

Special Thanks

We would like to express our great appreciation to Takeshi Azuma,
Director of the Paper Museum, members of the museum, All Japan
Handmade Papermaker's Association, and other people who helped us.

Ozu Washi Editorial Department

Please contact to Ozu Washi Editorial Department (+81-(0)3-3662-1184)
for your inquiry.

和紙　Washi　和紙ってなに？ What's Washi?

2020 年 1 月 6 日　初版発行

編著者	小津和紙編纂室　https://www.ozuwashi.net
発行者	原　雅久
発行所	株式会社　朝日出版社
	101-0065 東京都千代田区西神田 3-3-5
	電話（03）3263-3321（代表）

ブックデザイン：越海辰夫（越海編集デザイン）
イラスト／ 1 章マンガ：稲森直嗣
3 章画：岡田潤
編　集：一瀬正廣、国保奈月（小津和紙編纂室／電話 03-3662-1184）
　　　　エーアンドエー株式会社
英文編集チーム：編集・伊佐間 梨華　翻訳・村瀬 智子　校正・Bradley Bierton
印　刷・製　本：協友印刷株式会社

©OZUWASHISHIRYOHENSANSHITSU 2020, Printed in Japan
ISBN978-4-255-01160-8 C0076